SULCATA TORTOISE

Comprehensive Guide To Sulcata Tortoises: Habitat, Care, Nutrition, Health, Breeding, Characteristics, Temperament, Enclosure, Substrate, Temperature, Lighting, And Behavior

Ethan Harry

Table of Contents

CHAPTER ONE 4

    INTRODUCTION TO SULCATA TORTOISES 4

    Overview And Natural Habitat 4

    Physical Characteristics 8

    Behavior And Temperament 11

    Importance In Ecosystem 14

CHAPTER TWO 19

    CHOOSING A SULCATA TORTOISE 19

    Selecting A Healthy Tortoise 19

    Age And Gender Considerations 23

    Reputable Breeders And Adoption Options 27

    Initial Health Check 30

CHAPTER THREE 34

    HOUSING AND HABITAT 34

    Enclosure Types And Sizes 34

    Substrate And Enclosure Furnishings 38

    Temperature And Humidity Requirements 42

    Lighting And Uvb Exposure 46

CHAPTER FOUR 50

    DIET AND NUTRITION 50

    Natural Diet In The Wild 50

    Recommended Diet In Captivity 53

    Supplements And Treats 56

    Common Dietary Issues 60

CHAPTER FIVE 64

HEALTH AND WELLNESS 64

Common Health Issues 64

Preventative Care 68

Recognizing And Treating Illnesses 71

Finding A Reptile Veterinarian 74

CHAPTER SIX 79

BREEDING AND REPRODUCTION 79

Mating Behavior 79

Nesting And Egg-Laying 82

Incubation And Hatching 86

Raising Hatchlings 91

CHAPTER SEVEN 95

BEHAVIOR AND ENRICHMENT 95

Understanding Sulcata Behavior 95

Providing Mental Stimulation 98

Handling And Socialization 103

Addressing Behavioral Problems 107

THE END 111

CHAPTER ONE

INTRODUCTION TO SULCATA TORTOISES

Overview And Natural Habitat

The Sulcata tortoise, also known as the African spurred tortoise (Centrochelys sulcata), is a remarkable reptile indigenous to the southern edge of the Sahara Desert in Africa. This region includes countries like Chad, Sudan, and Ethiopia, where the climate is predominantly hot and arid, providing an ideal environment for these tortoises. Sulcata tortoises are exceptionally adapted to these harsh conditions, capable of thriving in temperatures that would be challenging for many other species.

In their natural habitat, Sulcata tortoises are commonly found in semi-arid grasslands, savannas, and scrublands. These landscapes offer a combination of sparse vegetation and open spaces, which suits their foraging habits. Their diet primarily consists of fibrous grasses and plants, which provide the necessary nutrients and hydration for their survival. The ability to extract moisture from their food is a critical adaptation, given the limited availability of water in their environment.

One of the most fascinating behaviors of the Sulcata tortoise is its proficiency in digging burrows. These burrows serve multiple purposes, the most important of which is to provide refuge from the extreme heat. Temperatures in the Sahara can soar to levels that are potentially lethal to many animals, but Sulcata tortoises mitigate this by retreating underground during the hottest parts of the day. Their burrows can be quite extensive, sometimes reaching depths of up to ten feet and lengths of up to thirty feet. By residing in these subterranean shelters, Sulcata tortoises can maintain a stable body temperature and avoid dehydration.

The construction of burrows also plays a crucial role in moisture retention. The arid environment of the Sahara is characterized by minimal rainfall, and the ability to retain moisture is vital for the tortoise's survival. The humidity within the burrows helps the tortoises conserve water, reducing the need for frequent drinking. This adaptation is essential for surviving in an ecosystem where water sources are scarce and often ephemeral.

Additionally, the burrows offer protection from predators. While the Sulcata tortoise's large size and hard shell provide significant defense, the burrows add an extra layer of security. By remaining hidden underground, the tortoises reduce their exposure to potential threats.

Overall, the Sulcata tortoise's ability to thrive in one of the world's most extreme environments is a testament to its remarkable adaptations. From their diet and foraging habits to their burrowing behavior, these tortoises have evolved a suite of strategies that enable them to survive and prosper in the challenging conditions of the southern Sahara Desert.

Physical Characteristics

Sulcata tortoises, also known as African spurred tortoises, are among the largest tortoises in the world. Adult males can weigh as much as 200 pounds, while females typically weigh slightly less. The carapace, or shell, of an adult Sulcata tortoise can exceed 30 inches in length. This shell is domed and rugged, with a bumpy texture and distinct growth rings. These rings are not only a unique feature but also help in estimating the age of the tortoise.

The legs of Sulcata tortoises are particularly notable for their robustness. Their thick, strong legs are well-adapted for both digging and supporting their substantial body weight. The front legs are equipped with large, overlapping scales that provide protection when the tortoise burrows. These scales act as a shield against the rough substrate and potential predators. In contrast, the rear legs are more column-like, offering a stable base to support their heavy bodies.

A distinctive characteristic of the Sulcata tortoise is the presence of spurs on their hind legs, which is where they derive their name. These spurs, though not entirely understood, are believed to assist in digging and may play a defensive role. The spurs can help the tortoise in creating burrows to escape the extreme heat of their native habitats, which include the southern edge of the Sahara Desert in Africa.

The heads of Sulcata tortoises are relatively large, with powerful jaws that are well-suited for their herbivorous diet. They possess beak-like mouths that enable them to tear and chew tough plant material, such as grasses and fibrous vegetation. Their diet requires strong, efficient jaws to process the food they consume. The positioning of their eyes on the sides of their heads provides a wide field of vision. This adaptation is beneficial for spotting predators and locating food sources in their expansive, arid environments.

Behavior And Temperament

Sulcata tortoises, renowned for their hardy and resilient nature, exhibit fascinating behavioral traits and temperaments. In their natural habitat, these tortoises spend a significant amount of time foraging for food and digging extensive burrows. Their diet primarily consists of various grasses, leaves, and other vegetation, which are rich in fiber and low in protein—an essential nutritional balance for their well-being.

As primarily herbivorous creatures, Sulcata tortoises thrive on a diet that supports their digestive health and overall vitality. They are solitary by nature, typically preferring to live alone rather than in groups. This solitary lifestyle is particularly evident in the wild, where they establish individual territories and seldom interact with other tortoises outside of the breeding season.

Male Sulcata tortoises, in particular, display territorial behavior, especially during the breeding season. They are known to engage in combative displays, ramming into each other with their shells in a bid to establish dominance and win the favor of females. These confrontations, while natural, are part of their instinctual drive to secure mating opportunities and assert their territory.

Despite their solitary and sometimes combative nature, Sulcata tortoises can develop strong bonds with their human caretakers. Over time, they can become remarkably tame and even show signs of affection towards their owners. Many Sulcata tortoises learn to recognize their caretakers and may respond when called, displaying a level of familiarity and comfort in their presence.

However, these tortoises also possess a streak of stubbornness and willfulness. This characteristic can present challenges in their care, requiring a consistent and patient approach from their owners. Proper handling and a well-structured care routine are crucial in managing their behavior and ensuring their health and happiness.

In captivity, providing an environment that mimics their natural habitat is essential for their well-being. This includes ample space for foraging and digging, as well as a diet that closely resembles what they would consume in the wild. Understanding their solitary nature and territorial instincts helps in managing their interactions, especially when housing multiple tortoises.

Importance In Ecosystem

Sulcata tortoises play a crucial role in their ecosystems, significantly impacting vegetation and supporting biodiversity.

As primary consumers, these herbivorous tortoises contribute to the control of plant growth. By grazing on grasses and other vegetation, they prevent overgrowth, ensuring that no single plant species dominates the landscape. This grazing activity helps maintain a balanced plant community, fostering a healthy and diverse ecosystem. Additionally, their selective feeding can influence the types of plants that thrive, contributing to the overall biodiversity of their habitat.

One of the most significant ecological contributions of Sulcata tortoises is their role in seed dispersal. As they forage, they consume various plant materials, including fruits and seeds. The seeds often pass through their digestive system intact and are later deposited in their droppings, spread across different areas of their habitat. This process helps in the propagation of various plant species, promoting plant diversity and aiding in the regeneration of vegetation. The dispersal of seeds by Sulcata tortoises ensures that plants can colonize new areas, enhancing the resilience and sustainability of the ecosystem.

Moreover, Sulcata tortoises are prolific burrowers. Their burrowing behavior is essential for their survival, providing them with refuge from extreme temperatures and predators. However, these burrows also have broader ecological implications. They create microhabitats that offer shelter and a cooler environment for a variety of other species. Insects, small mammals, reptiles, and even birds can use these burrows as hiding spots or nesting sites. The presence of such burrows can significantly increase the biodiversity of the area, offering a safe haven for organisms that might otherwise struggle to survive in the harsh desert conditions. The creation of these microhabitats by Sulcata tortoises exemplifies the interconnectedness of ecosystem members. The burrows help maintain a stable microclimate, which is essential for the survival of various organisms, particularly during extreme weather conditions. In this way, Sulcata tortoises indirectly support the survival of numerous species, contributing to the overall health and stability of their ecosystem.

☐

CHAPTER TWO
CHOOSING A SULCATA TORTOISE

Selecting A Healthy Tortoise

Choosing a healthy Sulcata tortoise is crucial for ensuring that your new pet has a long and healthy life. Here are the key indicators to look for when selecting a Sulcata tortoise:

Eyes

One of the first things to examine is the tortoise's eyes. Healthy Sulcata tortoises have bright, clear eyes without any signs of discharge or swelling. Eyes should not appear sunken or cloudy, as these can be signs of dehydration or illness.

Shell

The shell is another critical aspect to inspect. A healthy Sulcata tortoise will have a firm, smooth shell without any soft spots, cracks, or irregularities. The shell should have a consistent shape and be free of any deformities. A soft shell can be a sign of metabolic bone disease, a common issue in tortoises caused by inadequate calcium or vitamin D3.

Activity Level

Observe the tortoise's activity level. Healthy Sulcata tortoises are generally alert and active. They should move around with purpose and show curiosity about their surroundings. If the tortoise appears lethargic, unresponsive, or spends an excessive amount of time hiding, this could indicate health problems or environmental stress.

Appetite

A good appetite is a strong indicator of a healthy tortoise. Ask the seller if you can observe the tortoise eating. Healthy Sulcata tortoises should eagerly consume their food, primarily consisting of high-fiber grasses and leafy greens. A lack of appetite can be a sign of illness or improper care conditions.

Breathing

Pay attention to the tortoise's breathing. Healthy tortoises should breathe quietly and easily without any wheezing, clicking, or bubbling sounds coming from their nose or mouth. Any abnormal respiratory noises can indicate respiratory infections, which are common in tortoises kept in suboptimal conditions.

Additional Considerations

Beyond these primary health indicators, there are other aspects to consider. Check the tortoise's skin for any signs of parasites, abrasions, or infections. The skin should appear smooth and free of any unusual bumps or lesions. Additionally, ask about the tortoise's history, including its diet, habitat conditions, and any previous health issues. Reputable breeders or sellers should provide this information and be willing to answer your questions.

Environmental Factors

Understanding the environmental conditions in which the tortoise has been kept is also important. Sulcata tortoises require specific humidity, temperature, and UVB lighting to stay healthy. Ensure that the tortoise has been housed in an appropriate environment that meets these needs.

Age And Gender Considerations

When selecting a Sulcata tortoise, both age and gender are crucial considerations, as they significantly influence the care requirements and compatibility of your pet.

Age:

Younger Sulcata tortoises, such as hatchlings or juveniles, are generally more affordable and more readily available. However, they require extensive care and attention, particularly during their first few years. Young tortoises are more vulnerable to health issues and environmental changes, necessitating a controlled habitat with appropriate humidity, temperature, and diet. This stage demands a commitment to daily care and monitoring to ensure proper growth and development.

In contrast, older Sulcata tortoises are typically more robust and resilient, having surpassed the delicate early years. These tortoises are less susceptible to common juvenile health issues and can adapt more easily to new environments. However, older tortoises often come with a higher price tag and can be more challenging to find. An advantage of choosing an older tortoise is that they have already reached their full size, providing a clearer understanding of the space and resources required to accommodate them. Given that Sulcata tortoises can grow quite large, having a full-grown tortoise allows for better planning of its habitat and care needs.

Gender:

Determining the gender of a young Sulcata tortoise can be challenging, as their sexual characteristics become more distinct with age. However, knowing the gender of your tortoise is important, especially if you plan to keep more than one.

Male Sulcata tortoises tend to grow larger than females and exhibit a concave plastron (the bottom part of their shell) and longer tails. These characteristics become more noticeable as the tortoise matures. Males are also known for their territorial behavior, often displaying aggression towards other males. This can lead to conflicts if multiple males are housed together, necessitating separate enclosures to prevent injury and stress.

On the other hand, female Sulcata tortoises are generally smaller, have a flatter plastron, and possess shorter tails. While females are less likely to display territorial aggression, they may still require careful consideration if housed with males. If you have both males and females, be prepared for potential breeding and the associated responsibilities, including managing eggs and hatchlings.

Reputable Breeders And Adoption Options

When choosing a Sulcata tortoise, whether through a breeder or adoption, it's crucial to prioritize reputable sources to ensure the well-being of the animal and your satisfaction as an owner.

Firstly, opting for a reputable breeder involves thorough research. Look for breeders with established credibility and positive customer feedback. A reputable breeder will offer references or reviews from previous clients, showcasing their experience and commitment to the species. They should possess deep knowledge about Sulcata tortoises, readily answering queries about care, diet, and habitat requirements. Moreover, they will provide comprehensive health records and insights into the tortoise's breeding background, ensuring transparency and accountability in their practices.

Alternatively, adoption presents a compassionate opportunity to provide a home for a tortoise in need. Many Sulcata tortoises end up in rescues or shelters due to their large size and extensive lifespan, which can overwhelm unprepared owners. Adopting from these organizations not only offers a second chance to these animals but also aligns with ethical considerations of animal welfare. Rescues specializing in reptiles can effectively match you with a tortoise that suits your circumstances and provide ongoing support and advice, fostering a successful adoption process.

Choosing between a breeder and adoption hinges on personal preferences and ethical stances. Breeders offer the advantage of obtaining a tortoise with documented lineage and specific attributes, ideal for those seeking particular traits or planning breeding programs. On the other hand, adoption promotes the humane treatment of animals already in need of caring homes, emphasizing responsibility and compassion in animal ownership.

Initial Health Check

Before bringing your new Sulcata tortoise into your home, it's crucial to prioritize their health with an initial examination conducted by a veterinarian specializing in reptiles. This step ensures that your tortoise starts off on the right foot in its new environment. Here's a comprehensive overview of what to expect during this initial health check:

Physical Examination: The veterinarian will begin by conducting a thorough physical examination of your Sulcata tortoise. This examination includes inspecting the tortoise's eyes, nose, mouth, and shell. By carefully examining these areas, the vet can identify any visible signs of illness, injury, or abnormalities. The shell, in particular, will be closely examined for any cracks, lesions, or irregularities that could indicate underlying health issues.

Fecal Test: A fecal examination is an essential diagnostic tool during the initial health check. The veterinarian will collect a fecal sample to test for the presence of internal parasites such as worms or protozoa. Detecting and treating these parasites early on is crucial for the health and well-being of your tortoise. Depending on the results, the vet will recommend an appropriate deworming protocol if necessary.

Weight and Measurements: The vet will accurately weigh your Sulcata tortoise and measure its shell dimensions. These measurements serve as baseline data to monitor your tortoise's growth and overall health over time. Changes in weight or shell size can indicate nutritional deficiencies, illness, or improper husbandry, prompting adjustments to their care regimen.

Diet and Care Advice: Proper diet and habitat setup are critical for the long-term health of Sulcata tortoises. The veterinarian will provide tailored advice on the ideal diet for your tortoise based on its age, size, and health status. This guidance typically includes recommendations on appropriate greens, vegetables, and occasional fruits that constitute a balanced diet. Additionally, the vet will discuss habitat requirements, including optimal temperature, humidity levels, and lighting conditions necessary for your tortoise to thrive.

General Health Discussion: Beyond the physical examination and diagnostic tests, the veterinarian will engage in a discussion with you about general health care for your Sulcata tortoise. This may cover topics such as handling techniques, signs of illness to watch for, and preventive care measures like regular check-ups and maintaining a clean habitat.

CHAPTER THREE
HOUSING AND HABITAT
Enclosure Types And Sizes

Choosing the appropriate enclosure for a Sulcata tortoise is crucial to ensuring its health and well-being. Sulcata tortoises are known for their impressive size and can grow quite large over time. Therefore, providing ample space is essential to accommodate their natural behaviors and growth requirements.

A minimum enclosure size recommended for a single adult Sulcata tortoise is typically 8 feet by 8 feet. However, larger dimensions are highly preferable if space permits. These tortoises thrive in environments where they can roam freely and exhibit natural behaviors such as digging and exploring. For outdoor enclosures, which are often recommended, the larger the pen, the better, as it allows the tortoise to move around more comfortably and engage in its natural behaviors.

Outdoor pens offer several advantages for Sulcata tortoises. Firstly, they provide access to natural sunlight, which is essential for their overall health. Exposure to sunlight enables them to synthesize vitamin D3, necessary for calcium absorption and bone health. Additionally, natural sunlight helps regulate their circadian rhythms, promoting more natural behavior patterns and better overall well-being.

When designing or choosing an enclosure for a Sulcata tortoise, the materials used should be safe and durable. Tortoises are known to be strong and may attempt to dig or push against enclosure walls. Therefore, sturdy materials such as solid wood or heavy-duty plastics are recommended to prevent escapes and ensure the tortoise's safety. Mesh or fencing should be secure and predator-proof if used in outdoor enclosures.

In addition to size and materials, the layout of the enclosure is also important. Providing varied terrain with gentle slopes, hiding spots such as rocks or logs, and shallow water dishes for soaking are beneficial. Sulcata tortoises enjoy exploring different textures and may appreciate areas with sand or soil where they can dig and burrow.

Maintaining proper temperature and humidity levels within the enclosure is crucial for the tortoise's health. While outdoor pens offer natural sunlight, they should also include shaded areas to prevent overheating during hot weather. Indoor enclosures should be equipped with heat lamps or ceramic heat emitters to provide warmth, especially during cooler periods or at night.

Regular cleaning and maintenance of the enclosure are essential to prevent the buildup of waste and bacteria. Spot cleaning should be done daily, with a thorough cleaning and disinfection performed periodically to ensure a clean and hygienic environment for the tortoise.

Substrate And Enclosure Furnishings

Creating a suitable habitat for a Sulcata tortoise involves careful consideration of both substrate and enclosure furnishings to ensure their health and well-being. These elements not only mimic their natural environment but also provide essential enrichment and security.

Substrate Selection: The substrate, or flooring material, plays a crucial role in replicating the Sulcata tortoise's natural habitat. A recommended substrate consists of a balanced mixture of soil, sand, and hay. This composition allows the tortoise to exhibit natural behaviors such as burrowing and digging, which are vital for their physical and mental stimulation. It's important to avoid substrates that are excessively rough or sharp, as these can potentially injure the tortoise's feet. Instead, a substrate that closely resembles the texture of their native environment helps maintain their comfort and health.

Enclosure Furnishings: In addition to a suitable substrate, the enclosure should include various furnishings to promote the tortoise's well-being. Providing shelters or hiding spots within the enclosure is crucial as it allows the tortoise to retreat and feel secure whenever necessary. These shelters can be created using rocks, logs, or commercially available hide boxes designed for reptiles. By offering hiding places, the tortoise can reduce stress and feel more at ease in its environment.

Enrichment Elements: Enrichment is essential for the overall health of Sulcata tortoises. Incorporating elements that mimic their natural habitat not only provides physical exercise but also stimulates their natural behaviors. Rocks of varying sizes can serve as climbing surfaces or basking spots, while logs or branches offer opportunities for exploration and enrichment. Live plants, carefully chosen to be non-toxic and safe for tortoises, not only add aesthetic value but also provide additional hiding places and can contribute to the enclosure's humidity levels.

Maintenance and Considerations: Regular maintenance of the enclosure is necessary to ensure the substrate remains clean and free of mold or bacteria buildup. Spot cleaning should be done regularly, and the substrate should be replaced periodically to maintain hygiene. It's also important to monitor the temperature and humidity levels within the enclosure to create a comfortable and healthy environment for the tortoise.

Temperature And Humidity Requirements

Sulcata tortoises require specific environmental conditions to thrive. These tortoises originate from the arid regions of sub-Saharan Africa, so replicating their natural habitat is crucial for their health and well-being.

Temperature Requirements:

Maintaining proper temperatures is essential for Sulcata tortoises. During the day, the enclosure should provide a gradient with temperatures ranging from 80-90°F (27-32°C) on the warmer side to no lower than 70°F (21°C) on the cooler side. This gradient allows the tortoises to regulate their body temperature effectively by moving between warmer and cooler areas as needed. Nighttime temperatures can drop slightly, but they should not fall below 65°F (18°C) to avoid stressing the tortoises.

To achieve these temperature ranges, using heat lamps or ceramic heat emitters can be effective. It's important to monitor temperatures regularly with reliable thermometers placed at different levels of the enclosure to ensure consistency throughout the day and night.

Humidity Requirements:

Despite being adapted to arid environments, Sulcata tortoises still require moderate humidity levels for optimal health. Aim for a relative humidity of about 40-50% within their enclosure. This level helps maintain hydration and prevents the tortoises' skin and shells from drying out excessively.

Proper ventilation is crucial to prevent humidity levels from becoming too high, which can lead to respiratory problems and shell rot. Good airflow within the enclosure can be achieved through strategically placed vents and ensuring that the enclosure materials allow for adequate air exchange.

Environmental Considerations:

In addition to temperature and humidity, other environmental factors contribute to the well-being of Sulcata tortoises. Providing a spacious enclosure with a substrate that allows for natural digging behaviors is beneficial. Tortoises often enjoy burrowing, so substrates like a mixture of soil, sand, and coco coir can mimic their natural habitat and encourage natural behaviors.

UVB lighting is essential for Sulcata tortoises as it helps them metabolize calcium and maintain healthy shell growth. Ensure that the UVB lighting covers a significant portion of the enclosure, especially the warmer basking areas where the tortoises spend much of their time.

Monitoring and Adjustments:

Regular monitoring of temperature, humidity, and other environmental conditions is crucial. Make adjustments as needed to maintain the recommended ranges, especially during seasonal changes or fluctuations in weather conditions.

Lighting And Uvb Exposure

Proper lighting plays a crucial role in the health and well-being of Sulcata tortoises, particularly in their ability to maintain essential physiological functions. Among these functions, the synthesis of vitamin D3 is paramount, as it directly influences calcium metabolism and contributes significantly to shell health.

Sulcata tortoises, native to the arid regions of Africa, have evolved under intense natural sunlight. Mimicking these conditions in captivity is essential for their overall health. This is achieved through the provision of either natural sunlight or artificial UVB lighting within their enclosure. UVB lighting, specifically designed for reptiles, emits rays that are vital for the tortoise's ability to synthesize vitamin D3 in its skin.

In practical terms, UVB bulbs should be strategically placed overhead in the tortoise's habitat. This placement ensures that the UVB rays adequately penetrate the enclosure, reaching the tortoise and enabling the synthesis of vitamin D3. It's recommended that the tortoise receives exposure to UVB light for approximately 10 to 12 hours per day. This duration mirrors the natural daylight cycle that the tortoise would experience in its native habitat, supporting its biological rhythms and metabolic processes. Without sufficient exposure to UVB light, Sulcata tortoises can develop serious health issues related to calcium deficiency. Vitamin D3 synthesized from UVB exposure facilitates the absorption of dietary calcium, which is crucial for maintaining strong and healthy bones and shells. Inadequate levels of vitamin D3 can lead to metabolic bone disease (MBD), a debilitating condition characterized by weakened bones, shell deformities, and other systemic health problems.

For indoor enclosures, where natural sunlight may be limited or unavailable,

UVB lighting becomes indispensable. UVB bulbs designed specifically for reptiles emit light in wavelengths that closely match those found in natural sunlight, ensuring that the tortoise receives the necessary UVB exposure for optimal health. Regular monitoring of UVB bulbs is essential, as their effectiveness diminishes over time. Bulbs should be replaced according to manufacturer recommendations to maintain consistent UVB output.

CHAPTER FOUR
DIET AND NUTRITION
Natural Diet In The Wild

In their native habitats, Sulcata tortoises, also known as African spurred tortoises, subsist primarily on a diverse array of grasses, weeds, and succulent plants. Their natural diet is centered around high-fiber vegetation, encompassing a broad spectrum of broadleaf plants, dandelions, clover, and various grass species. This dietary diversity is essential for their overall health and well-being, as it provides the necessary roughage crucial for maintaining optimal digestive function.

Sulcata tortoises are adapted to a grazing lifestyle, spending a significant portion of their day foraging for food. This continuous grazing behavior not only ensures they meet their nutritional requirements but also helps in wearing down their beaks and maintaining proper dental health. Their feeding habits are intricately linked to the seasonal availability of different plant species in their environment, leading to variations in their diet throughout the year.

During the wet season, when vegetation is lush and abundant, Sulcata tortoises typically consume a variety of fresh, tender grasses and herbs. These plants not only provide hydration but also essential vitamins and minerals crucial for their growth and metabolic functions. As the dry season sets in and vegetation becomes sparse, their diet may shift towards more drought-resistant plants and fibrous grasses that offer sustained energy and nutrients despite reduced water content.

The high-fiber content in their natural diet plays a pivotal role in their digestive health. Grasses and fibrous plants require extensive chewing, promoting saliva production and aiding in the breakdown of tough plant materials. This process helps prevent digestive issues such as impaction and ensures efficient nutrient absorption from their food.

Foraging behavior in the wild also exposes Sulcata tortoises to a range of plant compounds, including phytochemicals with potential health benefits. These compounds may act as antioxidants or provide other physiological advantages, contributing to their overall robustness and resilience in their natural environment.

Recommended Diet In Captivity

When maintaining Sulcata tortoises in captivity, replicating their natural diet is essential for ensuring their health and overall well-being. The staple of their diet should primarily consist of a variety of grasses and leafy greens. These components are rich sources of fiber and essential nutrients, closely mirroring what they would consume in their wild habitats. By offering a diverse array of greens such as dandelion greens, endive, mustard greens, and collard greens, tortoise keepers can help ensure a balanced nutritional intake.

In the wild, Sulcata tortoises are opportunistic herbivores, grazing on a wide range of vegetation. To mimic this diversity, it is crucial to provide a rotational diet of different greens. This practice not only meets their nutritional needs but also keeps them engaged and satisfied, reducing the likelihood of dietary deficiencies.

While greens form the bulk of their diet, occasional treats in the form of fruits can be offered. However, it's vital to exercise caution with fruits due to their higher sugar content. Suitable fruits for

Sulcata tortoises include strawberries and apples, which can be given in moderation. Ideally, treats should not exceed 10% of their overall diet to prevent issues like obesity and metabolic disorders.

Aside from fresh greens and occasional fruits, access to clean water is fundamental for Sulcata tortoises. They require regular hydration, which can be supported by providing a shallow dish of water that allows them to soak and drink as needed.

When planning the diet for Sulcata tortoises, it's important to avoid certain foods that can be harmful. Foods high in oxalates, such as spinach and rhubarb, should be limited as they can interfere with calcium absorption and contribute to health issues like metabolic bone disease. Additionally, processed foods, dairy products, and meats should be strictly avoided as they are not part of their natural diet and can lead to digestive problems.

Supplements And Treats

Supplements play a crucial role in maintaining the health and well-being of Sulcata tortoises, ensuring they receive essential vitamins and minerals necessary for their growth and overall vitality. Among these nutrients, calcium and vitamin D3 are particularly vital, contributing significantly to the development and maintenance of their shell and bone health.

Calcium is essential for Sulcata tortoises as it helps in the formation and maintenance of their strong shells and bones. In their natural habitat, these tortoises would obtain calcium from the variety of vegetation they consume, but in captivity, it's essential to supplement their diet to ensure they receive an adequate amount. Vitamin D3, on the other hand, facilitates the absorption of calcium from their diet, playing a critical role in their overall calcium metabolism. These supplements are typically provided in the form of powders, which are sprinkled over their food several times a week. It's crucial to follow the manufacturer's guidelines for dosage to prevent under-supplementation, which can lead to health issues like metabolic bone disease, or over-supplementation, which can be equally harmful.

When considering treats for Sulcata tortoises, moderation is key. While small amounts of fruits can serve as occasional treats, they should not replace the primary diet of greens and grasses. Fruits like strawberries, mangoes, and melons can be offered sparingly to add variety to their diet and to serve as rewards during training or to stimulate appetite. However, fruits should not exceed more than 5% to 10% of their total diet to avoid disrupting the nutritional balance required for their optimal health.

The primary diet of Sulcata tortoises should consist mainly of dark, leafy greens such as dandelion greens, collard greens, and mustard greens, supplemented with other safe vegetables and occasional grasses. This high-fiber, low-protein diet mirrors their natural foraging habits and helps to maintain their digestive health.

In addition to supplements and treats, ensuring proper hydration is essential for Sulcata tortoises. They require access to fresh, clean water at all times, provided in a shallow dish that allows easy access for drinking and soaking.

By maintaining a balanced diet that includes appropriate supplements, occasional treats, and ample hydration,

Sulcata tortoise owners can promote their pets' longevity and well-being. Regular veterinary check-ups can also help ensure that their dietary and health needs are being met adequately, allowing them to thrive in captivity as they would in the wild.

Common Dietary Issues

Common dietary issues can significantly impact the health of Sulcata tortoises kept in captivity, necessitating careful management and attention from owners. One prevalent concern is the overfeeding of fruits or foods high in carbohydrates and sugars. This dietary imbalance can lead to several health complications, including obesity, shell deformities, and metabolic disorders. To mitigate these risks, it is crucial for tortoise owners to closely monitor their diet and ensure that the majority of their food intake consists of appropriate greens and grasses. These fibrous plant materials provide essential nutrients while helping to maintain proper digestive function and overall health.

Another critical issue faced by Sulcata tortoises in captivity is the lack of adequate supplementation, particularly with calcium and vitamin D3. Calcium is essential for the development and maintenance of their shells and overall skeletal health. Vitamin D3 is necessary for calcium absorption and utilization. Without sufficient levels of these nutrients, tortoises can develop severe health problems such as shell deformities and metabolic bone disease. Therefore, owners should supplement their tortoises' diets with calcium and vitamin D3 to ensure they receive the necessary nutrients for optimal health.

Hydration is also a vital aspect of tortoise care that can affect their digestion and overall well-being. In dry environments or during hot weather, Sulcata tortoises may struggle to maintain adequate hydration levels. To prevent dehydration, owners should provide a shallow dish of water for drinking and offer regular soakings. Soaking not only helps to hydrate the tortoise but also assists in softening their food, making it easier to digest.

Proper hydration supports healthy kidney function and aids in the elimination of waste products from the body.

CHAPTER FIVE
HEALTH AND WELLNESS
Common Health Issues

Sulcata tortoises, like any pet, can encounter specific health challenges that require careful attention from their owners. Understanding these common issues is crucial for providing optimal care and ensuring the well-being of these fascinating reptiles.

Respiratory infections are among the primary health concerns for Sulcata tortoises. These infections often arise from incorrect temperature or humidity levels within their habitat. Symptoms may include wheezing, nasal discharge, and lethargy. Proper monitoring and maintenance of environmental conditions are essential preventive measures.

Another prevalent issue is shell rot, a condition that results from improper humidity or inadequate hygiene practices. Shell rot can lead to bacterial or fungal infections on the tortoise's shell, potentially causing significant discomfort and health complications. Regular inspections of the tortoise's shell and ensuring a clean and appropriately humid habitat are critical for preventing this condition.

Metabolic Bone Disease (MBD) is a serious concern for Sulcata tortoises, stemming from deficiencies in calcium or vitamin D3. This condition manifests as weakened bones and shell deformities, compromising the tortoise's mobility and overall health. MBD is preventable through a balanced diet rich in calcium and exposure to proper UVB lighting, which aids in calcium metabolism and prevents bone disorders.

Parasitic infections are also a risk for Sulcata tortoises, both internally and externally. Internal parasites such as worms and external parasites like mites can adversely affect their health. Regular fecal examinations by a veterinarian and meticulous habitat cleanliness are essential practices to mitigate the risk of parasite infestations and maintain the tortoise's health.

In addition to these specific health concerns, general aspects of care play a crucial role in ensuring the well-being of Sulcata tortoises. Providing a spacious and enriched habitat that mimics their natural environment, including proper substrate and hiding spots, promotes their physical and psychological health. A balanced diet consisting of leafy greens, vegetables, and occasional fruits, supplemented with calcium and vitamins as needed, supports their nutritional requirements.

Regular veterinary check-ups are imperative for early detection of health issues and ensuring timely intervention. Veterinarians experienced with reptiles can provide guidance on preventive care, habitat management, and specific health concerns tailored to Sulcata tortoises.

Preventative Care

Preventative care is crucial for maintaining the health and well-being of your Sulcata tortoise. These measures focus on creating an environment that supports their natural behaviors and physiological needs, ultimately promoting their longevity and quality of life.

Habitat Maintenance: Ensuring proper habitat conditions is paramount. Sulcata tortoises require a spacious enclosure with adequate temperature gradients and humidity levels. The ambient temperature should range between 75-85°F (24-29°C) during the day, with a slight drop at night. Providing a basking spot under a UVB light source is essential for the synthesis of vitamin D3, crucial for calcium absorption and shell health.

Dietary Needs: A varied diet is essential to meet their nutritional requirements. Offer a combination of leafy greens such as kale, collard greens, and dandelion greens, alongside vegetables like carrots, squash, and bell peppers. Occasional fruits like strawberries or melons can be offered as treats but should not dominate their diet due to their high sugar content. Calcium supplements may be necessary to prevent metabolic bone disease, a common issue in reptiles lacking sufficient calcium intake.

Hygiene: Maintaining cleanliness in the enclosure is vital for preventing bacterial and fungal infections. Regularly clean substrate materials like hay or coconut coir, and provide a shallow water dish for drinking and soaking. During these cleanings, inspect your tortoise for any signs of illness such as nasal discharge, swollen eyes, or abnormal behavior.

Environmental Enrichment: To stimulate natural behaviors, provide a habitat that mimics their native environment. Use substrates like sand or a mix of soil and sand to allow for digging and burrowing. Incorporate hiding spots such as logs or clay pots to offer security and privacy. Creating opportunities for basking under a heat lamp and grazing on vegetation will encourage natural behaviors and promote physical activity.

Recognizing And Treating Illnesses

Detecting and addressing illnesses in tortoises is pivotal for their well-being, requiring keen observation and proactive veterinary care.

Firstly, vigilance is key in spotting signs of illness early. Changes in appetite, activity levels, shell appearance, and breathing patterns can indicate underlying health issues. Any deviation from normal behavior warrants immediate attention and closer inspection. For instance, reduced appetite or lethargy might signal a metabolic problem or infection, while changes in shell texture or color could suggest shell rot or nutritional deficiencies. Similarly, irregular breathing patterns may indicate respiratory infections or other respiratory distress.

When signs of illness manifest, seeking veterinary care promptly is crucial. It's essential to find a reptile veterinarian experienced in treating tortoises. These specialists can conduct diagnostic tests, such as blood work or imaging, to accurately diagnose the underlying condition. Based on these findings, they can recommend appropriate treatments tailored to the tortoise's specific needs.

Home care plays a pivotal role post-diagnosis. Following veterinary instructions meticulously ensures effective treatment. This may involve administering medications, adjusting habitat conditions, or altering diet to support recovery. For instance, if a tortoise is diagnosed with a bacterial infection, antibiotics may be prescribed, alongside recommendations for maintaining optimal humidity and temperature levels in their enclosure to promote healing.

Additionally, preventive measures are crucial in maintaining tortoise health. Regular veterinary check-ups, proper nutrition, and a clean habitat help prevent many illnesses. Adequate UVB exposure and a balanced diet rich in calcium and other essential nutrients support overall health and prevent conditions like metabolic bone disease.

Finding A Reptile Veterinarian

Finding a suitable veterinarian for your Sulcata tortoise is crucial to ensuring its health and well-being. Here are some essential steps and considerations to help you find the right reptile veterinarian:

Firstly, research and recommendations play a pivotal role. Reach out to fellow reptile owners, breeders, or local reptile clubs for recommendations. Their firsthand experiences can provide valuable insights into veterinarians who have expertise in treating tortoises like the Sulcata. Additionally, online forums and social media groups dedicated to reptile care often feature discussions and recommendations regarding reliable veterinarians.

Secondly, specialized knowledge is paramount. A qualified reptile veterinarian should possess a deep understanding of the unique anatomy, physiology, and common health issues specific to tortoises. This includes knowledge about proper nutrition, habitat requirements, and potential diseases that may affect tortoises, such as respiratory infections or shell problems. When researching potential veterinarians, inquire about their specific experience with treating tortoises and reptiles in general.

Thirdly, consider emergency preparedness. It's essential to know in advance where to seek help in case of emergencies. Not all veterinarians offer 24/7 emergency care for reptiles, so familiarize yourself with nearby emergency clinics or veterinary practices that specialize in exotic pets. Keep their contact information readily accessible, along with directions to their facility, particularly if you live in an area where specialized exotic animal care is not widely available.

When evaluating potential veterinarians, schedule a consultation or initial visit. This allows you to assess their facilities, observe how they interact with reptiles, and discuss your tortoise's specific needs. During this meeting, inquire about their approach to preventive care, their familiarity with tortoise health management, and the diagnostic and treatment options available for reptiles.

Furthermore, continuing education is a positive indicator of a veterinarian's commitment to staying updated with advances in reptile medicine. Ask about their participation in conferences, workshops, or specialized training related to exotic animal care. Veterinarians who actively seek to expand their knowledge and skills are more likely to provide optimal care for your tortoise.

Lastly, trust your instincts and establish a rapport with the veterinarian. Effective communication and a mutual understanding of your tortoise's care requirements are essential for a successful veterinarian-client relationship. Ensure that you feel comfortable asking questions and discussing concerns regarding your tortoise's health and well-being.

☐

CHAPTER SIX

BREEDING AND REPRODUCTION

Mating Behavior

The mating behavior of Sulcata tortoises, distinguished by their large size and robust build, is a fascinating display of natural instinct and adaptation. Typically occurring during the rainy season when food sources are plentiful and environmental conditions are optimal, these tortoises engage in a ritualistic courtship that highlights their social dynamics and reproductive strategies.

Male Sulcata tortoises are known for their aggressive interactions during mating season, often engaging in behaviors such as head bobbing, ramming, and occasionally biting. These actions are crucial in establishing dominance hierarchies among males, determining access to potential mates. Such displays are not merely combative but serve as a means for males to assert their superiority in the competition for breeding rights.

When a male identifies a receptive female, the courtship ritual begins with cautious approaches and gentle behaviors. The male may gently bite or nudge the female, testing her receptiveness and signaling his intent. This initial interaction sets the stage for further engagement, as the male seeks to ensure the female's willingness to mate.

The mating process itself is a methodical affair, marked by the male mounting the female from behind. This positioning allows for the alignment of their cloacas, the dual-purpose openings used for both reproduction and waste elimination. Cloacal alignment is essential for successful copulation, facilitating the transfer of sperm from the male to the female.

Copulation duration can vary significantly, lasting anywhere from several minutes to well over an hour depending on various factors such as the individuals involved and environmental conditions. During this time, the male maintains a firm grip to ensure proper alignment and effective sperm transfer, while the female typically remains relatively passive, allowing the process to unfold.

Beyond the physical mechanics of mating, these interactions underscore the evolutionary adaptations of Sulcata tortoises to their habitat and social structure. Mating behavior is not only a means of reproduction but also a manifestation of complex social behaviors and hierarchical dynamics within the species.

Nesting And Egg-Laying

After mating, the female Sulcata tortoise undergoes a critical phase in her reproductive cycle: nesting and egg-laying. This process is meticulous and vital for the survival of her offspring.

Following mating, which typically occurs during the breeding season when conditions are favorable, the female tortoise begins preparing for egg-laying by seeking out an appropriate nesting site. These sites are carefully chosen, often characterized by well-drained soil or sandy areas that facilitate easy excavation. This choice is crucial as it impacts the nest's stability and the incubation conditions necessary for the eggs' development.

The nesting behavior itself is a deliberate process that unfolds over several weeks post-mating. During this period, the female tortoise internally matures and develops the eggs, ensuring they are ready for deposition. Once she identifies a suitable location, she proceeds to dig a nest cavity using her powerful hind legs. This cavity serves as the protective chamber where the eggs will be laid and subsequently incubated.

When the nest cavity is sufficiently prepared, the female begins the egg-laying process. Clutch sizes can vary significantly among Sulcata tortoises, typically ranging from 15 to 30 eggs per clutch, although individual variations are not uncommon. Each egg is carefully laid into the nest cavity, with the female employing precise movements to position them securely. This process ensures that each egg is positioned optimally for incubation, promoting uniform development.

After all eggs are laid, the female covers them gently with soil using her hind legs. This covering is not just for protection but also helps to regulate the temperature and humidity within the nest, crucial factors influencing the development and hatching success of the eggs. The depth at which the eggs are buried and the ambient temperature of the nesting site are critical variables that directly affect incubation duration and the health of the developing embryos.

Throughout this entire nesting and egg-laying process, the female Sulcata tortoise exhibits instinctual behaviors honed over evolutionary time. These behaviors maximize the chances of survival for her offspring, ensuring they have the best possible start in life. As such, the meticulous care taken in selecting a nesting site, digging the nest cavity, laying the eggs, and covering them with soil underscores the maternal investment and biological strategies employed by these tortoises to perpetuate their species in often challenging environments.

Incubation And Hatching

Incubation and hatching are critical phases in the lifecycle of Sulcata tortoises, involving careful management of environmental conditions to ensure successful development from egg to hatchling. In their natural habitat, the incubation process begins immediately after the female tortoise lays her eggs. Unlike many mammals, which maintain consistent internal temperatures, reptiles like Sulcata tortoises rely on external environmental temperatures to dictate the pace of egg development. Warmer temperatures generally accelerate incubation, while cooler conditions prolong the process.

Captivity necessitates a controlled approach to incubation to mimic the natural environment. Breeders often utilize artificial incubators specifically designed to maintain optimal conditions. These devices regulate both temperature and humidity levels, crucial factors that directly influence the health and viability of developing embryos.

Monitoring these parameters closely throughout the incubation period is essential, as even minor fluctuations can significantly impact the outcomes.

The average incubation period for Sulcata tortoise eggs spans approximately 75 to 100 days under typical conditions. However, this timeframe can vary based on several factors, including the specific temperature and humidity levels maintained within the incubator. Optimal temperatures for incubation generally range between 82°F to 88°F (28°C to 31°C), with relative humidity levels ideally set around 80%. These conditions aim to replicate the warm, moist environments tortoise eggs experience in their native habitats, facilitating proper embryonic development.

During incubation, breeders meticulously observe the eggs, ensuring that environmental conditions remain stable. This diligence is crucial as any deviations could compromise the health of developing embryos or lead to unsuccessful hatching. Fluctuations in temperature or humidity beyond recommended ranges may result in developmental abnormalities or even mortality among embryos.

Maintaining proper humidity levels is particularly vital as it prevents dehydration of the eggs, which could otherwise prove fatal to the embryos.

Adequate ventilation within the incubator helps regulate moisture levels, preventing the buildup of excess condensation that might promote bacterial growth and compromise egg integrity.

As the incubation period progresses, breeders often use candling—a method involving shining a light through the egg—to monitor embryonic development. This technique allows them to track the growth of the embryo and assess its viability without disturbing the delicate incubation environment unnecessarily.

Ultimately, successful incubation culminates in the hatching of healthy Sulcata tortoise hatchlings. Once fully developed, hatchlings typically emerge from their eggs using a specialized egg tooth—a temporary structure that aids in breaking through the eggshell. Following hatching, newly emerged tortoises may require a period of rest and acclimatization before being introduced to their rearing environment.

Raising Hatchlings

Raising hatchlings, particularly of Sulcata tortoises, involves delicate care and attention from the moment they begin to hatch. These small tortoises emerge from their eggs using an egg tooth, a specialized projection on their beak that helps them break through the tough eggshell. This process marks the beginning of their journey, where they are particularly fragile and in need of nurturing care.

To ensure the well-being of hatchlings, breeders and caretakers create a warm and humid environment that mimics their natural habitat. This environment is crucial as it helps regulate their body temperature and maintains humidity levels essential for their health and development. Hatchlings are also provided with shallow water dishes to ensure they stay hydrated, as well as small, easily digestible food items such as leafy greens and finely chopped vegetables. These foods are essential for providing the necessary nutrients for their early growth stages.

In the initial years of their life, hatchlings undergo rapid growth, which necessitates adequate space to move and explore. UVB lighting is crucial during this stage as it aids in the absorption of calcium, essential for the development of their shells and bones. A varied diet is also paramount, as it ensures they receive a balanced intake of vitamins and minerals necessary for healthy growth and development.

Careful monitoring of hatchlings is essential to detect any health issues early on. Common issues include shell deformities or infections, which can be managed with prompt veterinary care and adjustments in their environment or diet. Breeders often keep detailed records of each hatchling's growth and health milestones to ensure they are progressing as expected.

As hatchlings mature, their care requirements evolve. They gradually become more resilient and less susceptible to environmental changes. However, proper husbandry practices remain crucial throughout their lives to ensure they continue to thrive.

CHAPTER SEVEN
BEHAVIOR AND ENRICHMENT
Understanding Sulcata Behavior

Sulcata tortoises captivate enthusiasts with their distinctive behaviors, rooted in their natural habitat and evolutionary adaptations. These herbivorous giants exhibit a range of behaviors essential to their well-being, making understanding their habits crucial for effective care in captivity.

As diurnal creatures, Sulcata tortoises are most active during the day, leveraging sunlight and heat lamps to regulate their body temperatures. This behavior mirrors their natural habitat in the arid regions of Africa, where they bask to absorb warmth and maintain optimal physiological functions. Providing adequate basking spots in captivity is essential, ensuring they can thermoregulate effectively.

One of the most notable behaviors of Sulcata tortoises is their affinity for digging. In the wild, they excavate burrows to escape extreme temperatures and evade predators, creating safe havens underground. This instinctual digging behavior serves both as a defense mechanism and a means to regulate their body temperature by accessing cooler underground environments during intense heat. In captivity, replicating this natural behavior is beneficial, achieved through providing a substrate that allows them to dig and burrow comfortably.

Feeding habits also reflect their natural environment. Sulcata tortoises are primarily herbivores, grazing on grasses, weeds, and other vegetation. Mimicking this diet in captivity with a variety of leafy greens, grasses, and occasional fruits ensures their nutritional needs are met, supporting overall health and vitality.

Social behaviors in Sulcata tortoises vary, with individuals often displaying territorial tendencies, especially males during breeding seasons. Understanding these behaviors helps in managing interactions between tortoises in captivity, preventing potential aggression or stress.

Observing and responding to these behaviors are crucial aspects of caring for Sulcata tortoises effectively. Providing a spacious enclosure with varied terrain, including areas for basking, digging, and grazing, promotes their natural behaviors and overall well-being. Regular observation allows caretakers to adjust environmental conditions and diet as needed, ensuring these fascinating creatures thrive in captivity as they would in their native African habitats.

Providing Mental Stimulation

Providing adequate mental stimulation is crucial for the well-being of Sulcata tortoises, despite their slow-moving nature. These tortoises benefit significantly from environmental enrichment, which can help keep them active and engaged in their surroundings.

Foraging opportunities play a pivotal role in simulating natural behaviors for Sulcata tortoises. By scattering food around their enclosure, caretakers encourage the tortoises to exhibit natural foraging instincts. This not only provides physical exercise but also mental stimulation as they search for and consume their food. Incorporating a variety of foods and placing them in different locations can further enhance this enrichment strategy, mimicking the diverse foraging challenges they would face in the wild.

Structures within the enclosure also contribute to mental stimulation. Logs, rocks, and low platforms create opportunities for climbing and exploration. Sulcata tortoises, despite their terrestrial nature, enjoy exploring different surfaces and heights. These structures not only diversify their physical environment but also prompt them to navigate and investigate their surroundings actively. Care should be taken to ensure these structures are stable and safe to prevent injury.

Toys and puzzle feeders are another effective method to engage Sulcata tortoises mentally. These can range from simple objects like balls or objects with interesting textures to more complex puzzle feeders that require effort to obtain food rewards. Such toys stimulate their minds by encouraging problem-solving behaviors and physical activity. Introducing new toys periodically prevents boredom and maintains their interest in exploring and interacting with their environment.

In addition to physical structures and toys, varying the enclosure's layout and introducing new objects periodically can prevent habituation and keep the tortoises mentally stimulated. Rotating items such as plants or rearranging structures can create novel challenges, prompting the tortoises to adapt and explore their environment differently.

Observing individual tortoise behavior can also guide enrichment efforts. Some tortoises may prefer certain types of enrichment over others, so offering a variety allows caretakers to cater to individual preferences. Monitoring their response to different enrichment activities can help refine the enrichment program to best meet their needs.

Overall, enriching the environment of Sulcata tortoises with foraging opportunities, diverse structures, and stimulating toys is essential for their mental and physical well-being. These activities not only mimic natural behaviors but also prevent boredom, encourage physical activity, and promote overall health in these fascinating reptiles. By providing a stimulating environment, caretakers can ensure that Sulcata tortoises thrive and lead fulfilling lives in captivity.

Handling And Socialization

Handling and socializing Sulcata tortoises requires a gentle and cautious approach, considering their natural tendencies and physical characteristics. While these tortoises are not inherently social creatures, they can learn to recognize their owners and tolerate gentle interactions over time. Here's how to handle and socialize Sulcata tortoises effectively:

Firstly, it's crucial to support their shell properly when lifting them. The shell is a fundamental part of their anatomy, providing both protection and structural support. When picking up a Sulcata tortoise, ensure you lift them gently and evenly to distribute their body weight. This minimizes stress on their skeletal structure and prevents potential injuries that could arise from mishandling.

Avoiding dropping a tortoise is essential for their safety and well-being. Dropping them, even from a low height, can cause significant stress and may result in injuries such as shell fractures or internal trauma. Handling should always be done with care and attention to detail, maintaining a secure grip throughout the process.

Limiting handling sessions is also advisable. While some tortoises may tolerate handling well, excessive interaction can stress them out. Sulcata tortoises, in particular, prefer minimal disturbance and may become anxious if handled too frequently or for extended periods. It's best to keep handling sessions brief and infrequent, allowing them ample time to rest and adjust to their environment without constant human intervention.

Despite their non-social nature, regular, gentle handling can acclimate Sulcata tortoises to human presence. This process involves slowly introducing them to handling routines while observing their reactions and comfort levels. Over time, they may become more accustomed to being touched and lifted, recognizing familiar caretakers and responding calmly during interactions.

When engaging with Sulcata tortoises, patience is key. Each tortoise may have its own tolerance levels and preferences regarding handling. Some may show more interest in human interaction than others, while some may prefer to retreat into their shells when approached. Understanding and respecting their individual behaviors can help create a positive and stress-free environment for them.

Addressing Behavioral Problems

Addressing behavioral problems in Sulcata tortoises is crucial for ensuring their well-being, as these issues often stem from environmental or care-related factors. Understanding and addressing these issues promptly can significantly improve the tortoise's quality of life.

One common behavioral issue observed in Sulcata tortoises is aggression, typically arising from territorial disputes exacerbated by small enclosure sizes. Inadequate space can lead to stress and heightened territorial behavior among tortoises. To mitigate this, it's essential to provide ample space within their enclosure. Creating separate areas or dividing the space can help in reducing confrontations between conflicting individuals. This approach allows each tortoise to establish its territory more comfortably, minimizing aggression.

Another prevalent issue is the refusal to eat, which can be attributed to various factors such as an incorrect diet composition or stress. Sulcata tortoises require a diet primarily consisting of leafy greens, grasses, and occasional fruits. Ensuring a balanced and varied diet is crucial to their nutritional needs and overall health. Stress can also play a significant role in their eating habits. Factors such as inadequate environmental conditions or changes in their surroundings can lead to stress-induced anorexia. Monitoring their feeding patterns and adjusting their diet as necessary can help in addressing this issue effectively.

Excessive hiding behavior is another indicator of potential stress or discomfort in Sulcata tortoises. This behavior often results from inadequate temperature regulation within the enclosure or insufficient hiding spots. Sulcata tortoises require a habitat that offers suitable temperature gradients, allowing them to thermoregulate effectively. Providing hiding places such as shelters or vegetation within the enclosure gives them opportunities to retreat and feel secure. Ensuring these environmental enrichments can significantly reduce stress-related behaviors like excessive hiding.

Addressing these behavioral issues requires a holistic approach that considers both environmental factors and the tortoise's specific needs. Regular monitoring of their behavior and environment is essential to identifying and resolving any emerging issues promptly. By providing adequate space, a balanced diet, and a well-regulated environment with appropriate hiding spots, caregivers can foster a healthier and more comfortable living environment for Sulcata tortoises.

THE END

www.ingramcontent.com/pod-product-compliance
Lightning Source LLC
Chambersburg PA
CBHW071834210526
45479CB00001B/135